AN OUTDOORSMAN'S

Greatest Discovery

FOR THE 21ST CENTURY

The Real Reasons Animals Are Detecting Your Sets and Devices

A book for science and the outdoorsman with an addition related to human application that will explain why the cases against Bayer-RoundUp should eventually be exonerated.

by **KIRK DEKALB**

with illustrations by Harvie Ann Cox

"There were testimonials from trappers that showed a four to one increase in catches of one trap/device over another with similar size, similar design, and almost identical materials. Many trappers using cage traps said they did not promote certain traps so as to limit competition in their area. They found what I found and observed but just didn't know why. All of the traps caught animals. It was only the best trappers that had two similar trap/devices, from two different manufacturers, that could tell the difference."

Concerning Bayer/Roundup:

"If proper testing is not done, the only evidence will be traces of a non-harmful chemical that got blamed for the occurrences of cancer. This text should be understood and its recommendations followed, or Bayer will continue the losses. It will, for many years, be blamed for the shortened lives of thousands of people if key points in this book are not more carefully researched and proven to be true."

- Kirk DeKalb

Table of Contents

Preface

Here are my findings based on infield testing, observation, real time video, research periodicals, scientific studies and reading trappers experiences with the devices mentioned. This was done over 23 years trapping professionally. During this time I experimented with body grip traps, round and standard, of various sizes. Snares and foot hold traps were used of several size and snare lock types. Testing was done with experience from over 120 different cage trap designs, many having multiple sizes in length, width, and height, as well as trigger design. They were square, round, rectangle, triangle, and pyramid in shape. This included doors that locked with rings, powered and not. Traps with powered saloon style doors on each end. Guillotine doors of various designs. Many had swing down powered lock bar doors combined with some traps that had at-least 3 different door designs on the same trap. Triggers on the cage traps used have been multiple numbers of pan designs to over 6 different wire trigger styles and type. Some were patented, several were not. Using trapping devices over 18,000 animals were caught, in the wild, and over 60,000 sets were made. During a 13 year period, of the 23 years, over 400 locations were set each year. More than one hundred thousand acres were accessed and over 14,000 animals were caught. In most of the 13 years the bulk of trapping was done, in less than 120 days, with a 24hr check.

95% of the work that was done removing animals was Animal Damage Control.

It wasn't until a magnetometer was purchased that the way the animals were responding to set traps and hunting/ trapping devices made complete sense. For almost four years I suspected, with very strong beliefs, but never could prove scientifically. I was always given doubt, from my findings, by respected professionals that I thought had the same experiences I had and the ability to interpret them.

What really got my attention were the testimonials from trappers that showed an up to four to one increase in catches of one trap/device over another with similar size, similar design, and almost identical materials'. Many trappers, using cage traps, said they did not promote certain traps so as to limit competition in their area. They found what I found and observed, but just didn't know they could use a magnetometer to tell which trap/device would be the best. All of the traps caught animals. It was only the best trappers that had two similar trap/devices, from two different manufacturers, that could tell the difference.

Concentrating on "the most effective traps", the only thing they "all" had in common was a decrease in magnetic field intensity and a larger percentage of adult animals being caught in the traps with less refusals.

Introduction

It has been proven that certain animals have a flavoprotein molecule sensitive to blue light called Cryptochrome. Cryptochrome may allow certain animals to sense, "feel", or see the magnetic field emitted by a trap or trapping device. Some animals with Cryptochrome have a better ability to detect the field than others with the same protein molecule. They may have the ability as a youngster to detect magnetic fields, but must learn from their mistakes to live to maturity. A one year old may not have had time to associate a magnetic field variance as danger. They must learn what sights, sounds, scents and other things related to a threat that would be harmful. That takes a multitude of experienced contacts over time. Many animals' brains don't mature for two to three years making it harder to retain and understand what has been learned until almost becoming an adult, especially if cells have been damaged during that time.

Foothold Traps

The way the trapping device or object is designed, materials used, placed, and temperature determines how animals can detect and react to it. For instance, a common foothold trap will emit a higher magnetic field reading at its dog and levers when set. The field is enhanced or "greater" if the trap has four coiled springs rather than having two coiled springs. (The same reason when an electric motor is wound) The magnetic field circles the jaws and gathers through the parts called the dog and levers. The electro magnetic field emitted from the trap at these locations can be 1/2 to 5 times higher at those points (*Image Index 1*) when the trap is set depending

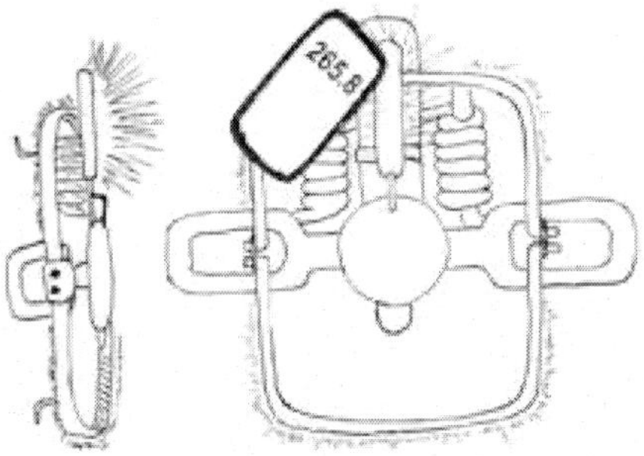

fig. 1 - magnetic field variance as recorded on foot hold trap 265.8μT on the trap/conditions at the time. I have found that when the set trap is bedded, predators will sense the magnetic field and paw at the highest magnetic emission point in the field created. When making a dirt hole set, by placing the trap dog toward the dirt hole and lure, the animal is less likely to dig from the back (side dog is attached) and is more apt to get caught.

Another way to avoid a dug foothold trap, by the animal, is to power bed the trap using nails. Not only does this firm up the trap but grounds the trap, to a degree, dramatically rearranging the intensity of the magnetic field away from the parts of the trap called a dog and levers. To do this nails are placed so as to make the trap more secure so the jaws are firm and can't move when stepped on. Placing the nails between the set jaws on each side as well as under the loose jaw seems to be best. In most cases, the magnetic field is changed and redirected so as to reduce what the animals targeted can sense. Power bedding allows enhanced walk through sets for fox and coyote. For the past one hundred plus years trappers have thought the digging up of a set trap was due to scent left by the trapper or a trap that moved when the jaw was stepped on. Though this played a factor, the detection of the magnetic field emitted from the trap in my opinion is primary with a predator known to utilize Cryptochrome in its system.

By placing the trap deeper and in more dense soil will allow the magnetic field variances to be less intensive, radiating from the trap. Also, by placing deep and at a steeper angle to the trap pan, the set trap will make it more difficult for the animal to dig without getting caught. This aligns the paw of the animal closer to the jaw's catch area when the trap fires.

Snares

When using snares and if the trapper is having "snare avoidance", reducing the generated field intensity is helped by using the smallest diameter cable possible for the animal targeted, as well as the snare lock itself. The larger the diameter of the set snare loop normally helps to reduce the magnetic intensity in the center of the snare. These last two sentences are especially true during a full moon. The moon has an effect of increasing the intensity of magnetic field allowing a snare or trapping device to be better seen and/or sensed by the animal. The magnetic field follows the circled snare cable intensifying at the snare lock and the snare swivel. A set 3/32 large washer lock snare's magnetic reading in the southern part of the United States can be as high as 4 times the average for the local average magnetic field (*Image Index 2 & 3*) taken at the washer lock and snare swivel. **Again, to reduce the magnetic field variance to an acceptable level for coyote the snare cable and lock size should be as small as possible and still function properly.** The end of the set snare at the swivel should be tied to where the radiating field it gives off would have less of a chance of being seen or sensed as a threat. [Electric current flowing in a circular loop will generate a magnetic field.] When an animal moves through the magnetic field electric current is generated. If the animal can sense the current he produces, moving through the field, he knows the magnetic field exists. The stronger the current the stronger the field the animal determines. He may not see the magnetic field but feel, or sense the current's strength to know it is there and whether to react.

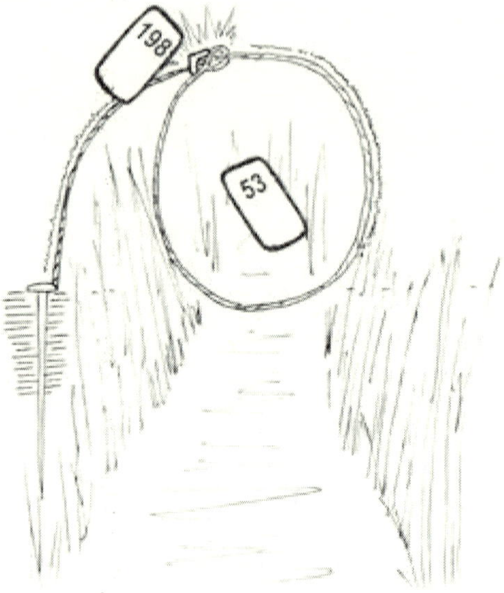

fig. 2 - magnetic field variance as recorded at the washer lock, 198μT. Magnetic field in center shown at 53μT, 15% over the magnetic field of the earth. To be most effective, this reading needs to be below the average magnetic field of the earth.

Body Grip Traps

When it comes to using 330 body grips they should be set completely under water, for best results. Locations that encourage the animal to be swimming at a fast pace are best. An example of this would be a location 10, or more feet down from a crossover. The animal will have time to build up speed, making it harder for him to change course or not commit when swimming toward and entering the trap. Stream current flow also seems to help, especially in warm weather conditions. Exposed devices are less avoided in cold water, or on wet rainy days when the wind and water is moving. The current of the water disrupts the sensing of any existing magnetic field's intensity by movement of molecules away from the device. The faster the flowing water around the encased trap, the harder it is to detect the EMF of the trap as the animal approaches. Testing shows that turning and setting the trap with the dog down sometimes helps to reduce a magnetic field the Cryptochrome armed animals can see or sense surrounding the jaws, when traps are exposed. The 330 traps that were tested had a much higher magnetic field reading on the corners, on the bend, at the top of the exposed trap (*Image Index 4*). Some traps and trap devices have **reduced** magnetic field readings (measured in micro tesla) going through the trap or device (*Image Index 5*). Some, depending on the manufacturing process and design, have **higher** magnetic field readings going through the trap. Highest magnetic field reading variances are found when the full moon is directly overhead and closest to the earth. A few hours can make a difference with field intensity effect of the moon at this stage. Some devices may show a reduced field in the morning (going through the device), but emit a higher reading that night when the moon is overhead and full. **There again testing has shown the trapper wants a reduced magnetic field intensity going through every trapping device and/or set area (when using footholds) for less avoidance when set is exposed.** Watch night video, when a beaver or otter first encounters a 330 set exposed he will go to one side, and then raises up, sensing the magnetic field. If the field is reduced in the inside- center of the trap he may back up and go through or go around and then, come back through the

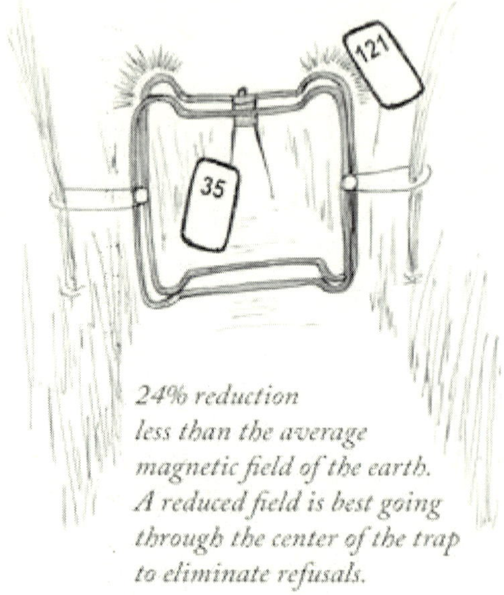

fig. 3 - magnetic field variance as recorded on body grip trap 121μT.

trap. If the average magnetic field reading is higher in the inside middle, going through the trap, he probably will avoid the exposed trap altogether. Many talk of animals becoming "square shy". That may be the case but an emitting magnetic field radiating from the exposed trap from what I have learned says, "I am here," from a distance to an animal carrying the Cryptochrome molecule. The negative response to magnetic intensity could have been learned first from a previous catch that the animal witnessed. It could have then sensed the magnetic field of the device on the caught animal as he investigated. When he recognized the magnetic field again he knew to associate it with danger. He may not be square shy but, magnetic intensity shy from that point on.

Cage Traps

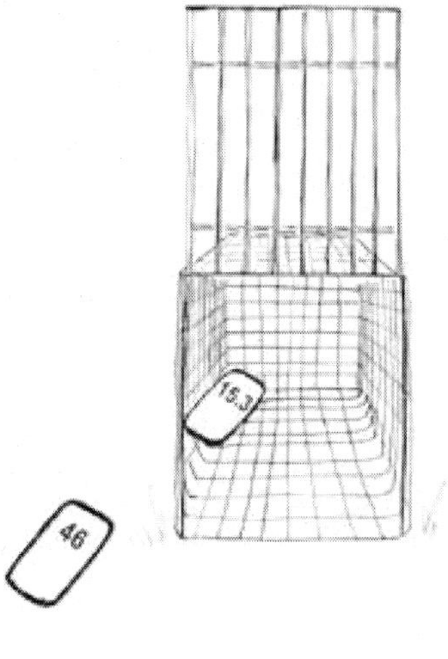

fig 4 - Magnetic field readings shown inside and outside of properly made single-door cage trap.

Cage traps are much like the large body grips in that they emit the magnetic field from the outside when set. If the internal magnetic field intensity decreases as the animal enters, negative ions are being released. Negative ions are calming to animals and refusals are lessened when animals are encountered (*Image Index 6*). How the cage trap is designed and placed is also very critical, as to refusals being reduced. The type of trigger and how it works can actually create a visual appearing barrier to some animals. This seems to be especially true of older animals with Cryptochrome that have learned from their mistakes they lived through in the past. A good example would be placing a double door cage trap with high magnetic internal variance, set as the only avenue for the animal to escape an attic. The animal, sensing a strong magnetic field inside the cage trap, panics, and creates his own escape exit avoiding the capture, which adds another hole in the attic. Another example would be double doored traps set over or in holes and the animals being trapped dig new exit holes to avoid the high magnetic field sensed in the trap. They may stay in the den or holes as long as lack of food and water allow, avoiding going through the trap, unless the trap is removed or a trap with a low and reduced interior magnetic field capability replaces it. I have found also the same situation can happen with body grips for the same reasons.

Some single door cage traps have a low magnetic reading as the trap is entered, but has increased readings over average at the pan or wire trigger. Others have a high intensity at the door. Both situations can cause refusals. Many cage traps with double doors tested have reduced magnetic field intensity on one end going through or entering the trap and a higher reading on the other end when the trap is entered. **There again what I have found is that the magnetic field reading should be a decreasing micro tesla on the inside no matter which way the animal enters, to reduce refusals.** (*Image Index 7*)

Some cage traps are designed to use live bait. How that bait is displayed determines the path the approaching predator takes in working the trap (*Image Index 8*). For instance if the trap is designed to have live bait in a protective compartment under the floor of the trap a retreating bird is just the ticket to draw the now circling, focused predator, into the reduced magnetic field inside the set trap.

CAGE TRAPS

No matter how well designed the trap is or any trapping device, a trapper should use all of the best tools in the box. The cage trap and all trapping devices emit an electro magnetic field. How the trapper is able to overcome the predators' senses is paramount to being successful. Using lures and baits in relation to the devices strengths and weakness coupled with the knowledge of the predator enhances the ability to catch.

Factors that determine magnetic field intensity

There are many other factors that affect the animals' ability to detect magnetic fields. The further north you go the magnetic field of the earth is greater. The further north you go the less affect the moon and sun effect the variance of the magnetic field, thus making the animal less likely to respond to changes in a negative way. Cold weather and reduced sunlight have a great bearing on magnetic variance. In a warm-moist climate the fields' intensity will actually increase the break down of molecules in placed lures and the human scent the trapper leaves, thus making the animal avoid the set because it enhances his ability to smell. On the other hand with a properly constructed cage trap/trap device the magnetic field reading is reduced inside the trap/ device allowing negative ions to flow reducing the speed of decay and breakdown of lures and bait. In the properly designed cages using live bait, flowing negative ions have a calming effect on the birds or animals used for bait. Included is a positive affect the negative ions have in promoting their good health while being held.

Detection of Magnetic Fields By Animals Having Little or No Cryptochrome

Cats are one animal not in the category of being able to sense magnetic fields because they don't have Cryptochrome. A good example of a cat's lack of detection is the use of footholds being effective and common in exposed sets for bobcats. That is why they should be limited in this discussion as some may not realize. What can and does happen with cats is their ability in conditions of ice and moisture to feel vibrations and/or hear a hum of the magnetic field given off by the snare cable when set, similar to the reason for the humming of power lines near your home. This takes place when the electrical field strength on the snares surface is larger than the intensity of the field needed to start the flowing of electrical current. The amount of the intensity of the discharge at the snare lock that causes the sensed vibrations can be affected by barometric pressure, water, ice, snow, as well as humid periods of morning fog. The cat may nuzzle the snare out of curiosity (feeling vibration and noticing a weak humming sound) and go around the first time, pushing the snare to the side. He was curious, just set the snare back. Feeling safe from the first pass, if the snare is set back in place, the cat will probably go through the set snare the next time he comes through, especially if weather conditions have changed. I have found the same to be true of beaver and otter. This also explains why some animals such as coyotes avoid snares in open runs, but will go through at fence crossings with a set snare. The snare is usually attached to the metal fence, thus changing how the magnetic field appears or now does not appear to the animal. Same may be accomplished by using a metal kill pole, where legal. This can eliminate and reduce his feeling a need to avoid the crossing leading to his capture.

Hunting Application

It was simply a matter of using the same testing methods to test hunting equipment. Testing metal battery operated feeders told the tale. I couldn't understand why it took so long for the animals to adjust to the metal feeders over a simple gravity pvc feeder. I thought it was the spinner doing the scaring, but the camera showed that the animals were avoiding what they saw and felt. The metal feeder took some getting use to by the wildlife. The low magnetic field pvc feeder was accepted much sooner and visited more often the first year they were used in a new area. The animals had to grow accustom to some of the new metal devices. The magnetometer explained which ones and why.

I learned metal deer stands had a high intensity that flowed up the rails and radiated at the top. The wooden ladder stands seemed to be less intrusive. It made since that a radiating field made the hunter easier to detect that varied with conditions and the magnetic intensity of equipment used. The ability to detect EMF's would only be evident in certain animals as described earlier in this text.

Negative Effects of Magnetic Fields and ROS

Recently more information has been found that a magnetic field, depending on its intensity and length of contact, can increase Reactive Oxygen Species (ROS), by causing Cryptochromes' to synthesis ROS. Too much ROS leads to damaged cells, reducing the animal's ability to use its senses and fight off infection. Too much synthesized ROS also leads to more genes that are responsive to ROS in a negative way. It may be possible that some animals can sense their body/ cells are being damaged by electro magnetism. With time they learn to avoid the magnetic fields that are harmful. It has been proven that dogs can detect cancer in humans and other animals. I believe coyotes can develop the ability to detect the harm that magnetic intensity's above the normal magnetic field of the earth can do to their body. I believe the abrupt reaction from mature coyotes when they encounter a magnetic field that is 25% or more above the average micro tesla reading for the location, is an example. Some cage traps that project a higher than average intensity at their doors, according to the videos I have seen, seems to all point that way. It is almost like a reaction from touching an electric fence. Fox will get a reaction, but not like a coyote.

Lying under a piece of discarded equipment that creates a high magnetic field, could be one way the coyote first encountered the feeling. A trap would not have to be the first object to emit a magnetic field encountered. The negative effect on cells could start in as short a time as a few days, according to what I have learned in the most recent published reports about cell damage. Interaction with other coyotes, over time, would increase a coyotes' ability to develop an understanding as to the need to avoid such areas and devices when found.

Temperature Effects

Information gathered for this text was evaluated to show temperature effects. Temperature averages for the 45 days each year of peak trap/device catches (In the Deep South) were analyzed from thousands of catches. This was done to find out what average daily temperature low had to be reached to maximize catch rates. Because testing was mainly done in the Deep South, this would also represent at what temperature was the minimum for determining the best trapping conditions for animal movement and magnetic field detection by animals. Of the 13 years evaluated two years (Test period-December 15 to January 31) had the highest catch rate for the 45 days evaluated. The average daily temperatures were 47 degrees with an average ground temperature at two inches was 50 degrees, during the most productive years. The years with lower catch rates, during the period, were only about 5 degrees higher average than those three year averages for the 45 days. This indicated to me that magnetic intensity in the metals in the devices used started indicating a variance intensity that was being detected in a negative way as the average temperatures rose above 50 degrees. This temperature average also showed 50 degrees or below was best for animal movement and activity. This explains why, in the Deep South the ability to catch and animal movement varies so much over a 150 mile area, some years from the next, caused by weather patterns. It is common for the temperature patterns to vary 5-7 degrees over 150 miles north to south. When average temperatures continued to rise above 55 degrees catch rates begin to drop dramatically, explained by the information in this text. It has been what I have experienced every year here beginning in March and lasting until about Thanksgiving in the Deep South. This period varies also from North to South by existing water temperature when water trapping. Water temperature in a Wisconsin stream in summer could be as much as 30 degrees colder than a Deep South stream on the same day. This would affect the intensity of the existing field of an exposed trap making it easier to catch an animal as has been discussed previously.

Hypothesis

The best trapping device will have a reduction of magnetic field intensity and a larger percentage of adult animals being caught in the device. It will have fewer refusals than its competition has because of that reduced magnetic field intensity. Up to a 4 to 1 increase in ability to catch, directly related to the magnetic reduction sensed by the animals working the set trap or trapping device.

When the magnetic field increase is generated from a human associated device, there is always a reason for the animal to be additionally cautious.

There is just so much that we are able to test with technology that wasn't available just a few years ago. These discussions will surely lead to much more. There is much more.

HYPOTHESIS

Summary

1. Plants and some animals can detect magnetic fields.
2. Some animals have a better ability to detect magnetic fields than other animals with the same magnetic sensing molecules.
3. The way the device is designed, material used, location, and how it is placed determines how the animals will react.
4. The magnetic field follows paths through the device which determines the strength and variance that allows the animal to detect where and how he will respond to the device.
5. Power bedding foot holds with nails can dramatically rearrange or redirect the magnetic field on some foothold traps.
6. To reduce snare avoidance use the smallest cable and lock possible needed for the animal targeted.
7. A full moon and directly over head being closest to the earth increases the ability of the animal for detection of device's because of the increase in the variation intensity of the magnetic field.
8. Some animals that do not carry Cryptochrome in their system

can still sense a magnetic field due to buzzing or humming like vibrations from the trapping device cause by the varying intensity of the magnetic field across parts of the device.

9. Body grip traps should be set completely under water, especially in warm months and warm Deep South environments.
10. Cage traps and body grip traps interior magnetic field can sometimes be reduced by basically turning the trap upside down.
11. Best if there is a reduced magnetic intensity detected through every trapping device and set area (when using footholds) for less avoidance when traps are exposed. A decreasing micro tesla reading should be experienced on entering the inside of a cage trap no matter which end is entered.
12. Positive ions produce negative affects. Negative ions produce positive affects.
13. The further north above the equator the higher the magnetic field of the earth and the less the magnetic field variance intensity.
14. Some animals detect the magnetic field by their ability to feel the current they make moving through the field. The stronger the field intensity the stronger the current. The easier it is to detect/sense with, or without the ability to see.
15. Using lures and baits in relation to the devices strengths and weakness coupled with the knowledge of the predator enhances the ability to catch.
16. The end of the set snare at the swivel should be tied to where the radiating field it gives would have less of a chance of being seen or detected as a threat. Exception might be when setting at crawl under and attaching to a fence or a metal kill pole.
17. Temperature evaluation showed best period to hunt and trap is when soil and air temperature daily averages (over a 45 day

period) are less than 50 degrees.

18. Even if an animal can not feel or see a high intensity magnetic field, where new devices are set and scent remained, the magnetic fields disruption of molecules will enhance detection by smell.

Deductions Made from My Findings and Relating Them to Humans

In the text of *An Outdoorsman's Greatest Discovery For The 21st Century* descriptions were given showing how magnetic fields were followed and detected in various devices. It showed how the flow of ions could be mapped with a Tesla meter. It also gave my hypothesis that cells could be detected, weakened, and damaged by magnetic fields as low as 25% above the average magnetic reading of the earth in the test areas. Taking what I learned I went further and applied what I learned from animals to how, where and why we as humans could be affected every day.

Let's use an example of a framing contractor:

He gets up every morning and gets in his car or truck. His truck emits a 50% higher than average magnetic field coming from his steering wheel. He gets to the job site and puts on his nail apron with a framing hammer and attaches his measuring tape to his belt at the hip. The measuring tape has a magnetic field projecting an intensity

fig. 5 - Construction worker shown with radiating tools.

which is 5 times greater than average of the earth where he is located. His hammer is giving about the same reading close to his body. He starts to do his work. It becomes necessary to pry up a 2x4 that was placed in the wrong location, so he grabs a crowbar. The crowbar has a magnetic intensity the same as the hammer and tape. The crow bar is not long enough to give the force needed to pry the 2x4, so he gets his pry bar out of the back of his truck. It has a magnetic reading that is higher than all of the tools he is carrying and has been using. One of his workers comes up and tells him of a leaking 4"pressure line. He reaches in his tool box and pulls out his largest pipe wrench. It has a magnetic reading far above all of the other tools. He uses the tools every day for 9 hours a day, five to six days a week. He also drives home and from site to site in the same truck. When he arrives home he relaxes in a chair on the patio. His wife brings him supper and places it on the patio table. Both the chair and table has a magnetic field of 4 times that which is average for the earth at that time. The tired contractor has developed a habit, for the last ten years, of devoting an hour each night searching on his I-Phone in bed. He plugs in both ear phones, so as not to disturb his wife beside him. Each side of his head is now receiving over 15 times the average of earth's magnetic field intensity. To add even more fuel to the fire, he lives in an area of the country known for its magnetic faults. There is a higher than average magnetic field radiation affecting his well water. The contractor has no clue what he is being subjected to for a full 24 hours each day, year after year.

You tell me this, if an animal as smart as a coyote can sense imminent danger with a magnetic field that is many times less than what that contractor is exposed to every day, for long hours at a time. What danger does that man possibly have to look forward to in the future, unless his life style changes? Some people have genetic ability to withstand infection more than others. What I have learned tells me if he gets cancer or has a major medical setback, it probably wasn't an initial chemical exposure. It was the magnetic field intensity caused by the tools he uses, chose to use, climate, products used, well water and location. The exposure would weaken his immune system every day, changing how cells respond/react, and set him up for a major setback. Stress could initiate it, or contact with something like the flu. It would not affect him as much, while he was young, but as he got older the constant magnetic intensity over time would increase his chances for major problems.

Here is another example of what I have learned. This one makes much sense and is kind of funny when you consider how we are manipulated and don't know it:

In my text of *An Outdoorsman's Greatest Discovery For The 21st Century* I explain how a double door cage trap having a lower magnetic reading, going through the interior of the trap has a calming effect on the encroaching animal. This is best if you are trying to be the most successful trapper. When you visit one of the major stores that carries hardware and other items, notice how the isles have racks holding merchandise on each side. Turn on your magnetometer. You will find that the EMF reading in the aisle between the racks is almost always less than the average magnetic field of the earth. The store layout has been designed to where negative ions that have a calming effect are in the center aisle moving toward each side. The ions are becoming positively charged as they are drawn up the racks holding merchandise. This is the same as in the interior of a correctly made cage trap. Now "you" are like the animal that is being targeted, this time by the store. Add a little soothing music and you are a shopper that is more apt to linger and buy more product as you walk down the aisle. (*Image Index 9*)

It all makes more sense now, doesn't it.

Thoughts for the Future

By describing the aisles of a store and comparing the layout to a properly made cage trap, we can deduce that some merchandise should be stored in the aisle. This might be tried to prevent the positive ions moving up the racks from attacking any organic items and fluids that would be susceptible to molecular breakdown stored on the racks. Would it be practical always? The answer would be an absolute NO. Changing the placement and type of product could change how the ions in the field flow, their intensity, and whether they are positive or negatively charged. That is where a magnetometer would come into play. In other words, it would tell you how harmful the magnetic field might be. From that, how the product would be displayed and placement would be determined.

The store probably just doesn't have an understanding of basic magnetic fields. Making the changes needed quickly, could change history. It would affect all stores. Structural, material, and other changes to design, if implemented too fast, could destroy the very progress we want to continue. How magnetic fields respond could just be a matter of product location related to what the store already uses. Understanding would be an on going ability learned over time and adjusted accordingly to each location as in the very reason this text was first written.

For example: Some grocery stores place vegetables and fruit on slanted displays. Due to some frame designs and where the produce is placed, the display actually pulls negative ions through the area presented. This increases the shelf life of the products displayed. Does the store know this is taking place? The store probably has found, over time, which displays they have been deploying work best and they just continue doing what they have done in the past. Understanding why, will allow the stores and work environments to make corrections on their own over time without costly regulation. This is what a magnetometer has taught me.

Feeding The World Could Be Devastating Without Bayer's Roundup/Glyphosate: Not being able to feed a fast increasing world population could be far more damaging than actions trying to protect the consumer through regulation. Sometimes it might be better to realize what has been taking place, and slowly adapt to the changes needed. A very good example of this last statement would be the devastating effect on our food supply without the availability of glyphosate (Round Up) to farmers through out the world.

Bayer/Monsanto with its research could be the ticket to simple solutions to many big problems in the near future. The piling on, of all current lawsuits, may hurt us all. It will slow down the desire and financial ability to develop products in the future that are protecting us. We should never forget what is meant by the saying: "Be careful not to bite the hand that feeds you". We would probably be best served by "nibbling" for a while instead of an all out attack, until we know more answers.

Side notes concerning Glyphosate: I had no "completely" detailed access, as to how tests were carried out, to prove or disprove the cases against Glyphosate. I did watch hours of testimony and read transcripts of testing procedures. I am not a scientist, but if the testing does not allow/consider magnetic field interference on a daily basis, the testing is not valid and should be thrown out. There is now enough scientific evidence to prove severe molecular and cell damage can be caused by daily contact of magnetic field intensities. Magnetic field intensity variance should be measured at every site, thoroughly, when animals are being used for testing. Micro tesla readings should begin "prior" to the birth of the animals used for testing. They should be continued at every location testing is carried on, through the end of analysis. I believe it is not so much the magnetic field intensity received from one source or the same amount every day. I believe it is various sources of radiation with varying intensity, hitting us at different times, as the day progresses. Over time the constant and inconsistent intensity effect on the cells in our bodies becomes harmful. It would be like a young child wiggling his tooth until it weakens enough to come out. All of the testing I have seen almost always is based on constants and deviation from those constants. Measuring exposure to magnetic intensity is different for every person and varies greatly through out the day, everyday, no matter the occupation. Just a short distance away from a radiating object would matter. Close contact time of radiating material and liquids should be short and avoided when possible. The animals are proving that to us by their avoidance of magnetic fields when using animal traps and devices.

How many laboratories are testing using varied exposure to

changing magnetic field intensities over a 24 hour period for years? (In micro tesla) The lab animal/mouse to be used for testing may have been subjected to harmful magnetic field radiation in its mother's womb before it was used for testing. This alone would invalidate a comprehensive test, whether later magnetic field testing was done or not. Remember, radiation not only alter cells but also the genes within those cells. The mice exposed to magnetic field intensity or radiation before birth could very well develop cancer without contact with any chemical due to damaged genes or cells.

In the first three suits (U.S.) against Roundup the persons bringing the suits all lived in California, which has numerous fault lines. Several labs doing Roundup testing are in California. Have the research labs in California done extensive testing using well water from California wells when testing mice? Did the plaintiffs' drink and use water from untested wells close to fault lines? A large number of wells in California have unsafe levels of dissolved solids probably caused by contact over time with positive ions from the EMF of the fault lines, or uranium deposits. Most landowners are responsible for determining the safety of their own wells. If untested (fault area) water was being used, the well water could have a very harmful effect on the applicator, the chemicals used, and all that come in contact with the radiated water over their life time. I don't believe that was considered or knew to be considered for the lawsuit. We normally "only" consider drinking water to be filtered. In a high fault area "all" well water that could be in contact, would need to be filtered and checked after being filtered, regularly. I would think the majority of the users of Roundup that used quantity got their mixing water from a well. One could go back and review tests from the wells over 30 years and results from wells near the test areas to check for abnormality. Even if it was filtered for harmful solids, reducing positive ions and normal water molecular structure should be a desired affect when testing.

Water/ chemical mixes, contaminated by sources of radiation would also be more easily absorbed by the skin especially in combination with a small amount of surfactant speeding up the process. The exposed person's cells would have traces of chemical even if all the cell damage was done from the positive ions of the radiation the water was exposed to. If proper testing was not done, the only evidence would be traces of a non-harmful chemical that got blamed for the damage.

We need to be certain of our findings before we cripple our food supply.

I never thought that my research would find me writing the text I have.

Don't give up when searching. I asked the head of the engineering department at one of Georgia's largest universities if animals could detect the magnetic field of a cage trap. He gave me a very sarcastic "No". He said an animal would not be able to sense a magnetic field that low. He acted as if I was wasting his time. To the professors' credit, he did call back. I left messages at two of the largest universities in the state and never got a call back.

What is proven in this book is basically that "all" traps/ devices are subject to being identified by some animals, because of the magnetic field they produce. Magnetic fields can be identified and their intensities measured, giving direction as to how to take advantage of, or avoid exposure. A magnetometer interprets what animals are telling us by showing us their response to what we cannot see. The animals that have the most genetic ability to confirm the harmful effects of EMF/radiation contact have been telling/warning us for over a hundred years.

Use this text to prove it for yourself.

Correlations Between Risk Factors for NHL and Findings in This Text

Non-Hodgkin's Lymphoma Risk Factors, *as cited from The National Cancer Society. *author's commentary seen in bold*

Age: Most cases NHL occurs in patients over the age of 60. **Day to day EMF exposure will normally take years to weaken cells enough to develop cancer.**

Gender: NHL is more likely in men than women. **This makes more sense with long term exposure to EMF's due to the activities and occupations of men.**

Race, Ethnicity, and Geographical History: Caucasians are more likely to develop NHL than Asians and African Americans. **Cell type and makeup allows for more tolerance of radiation that might lead to developing NHL in some people. From my experience as a contractor in the Southern United States, African Americans can tolerate higher exposure to the sun/heat than Caucasian Americans.**

CORRELATIONS BETWEEN RISK FACTORS FOR NHL AND FINDINGS IN THIS TEXT

Family History: First degree history to a relative is a factor. **The closest relative would be more likely to be exposed to the same environment as the person that developed NHL. That relative would be more likely to have experienced the same EMF levels that his father did.**

Exposure to Chemicals and Drugs: Studies have suggested chemicals may be linked to increased risk of NHL. **EMF would not have the same effect over time as chemical exposure unless it was a daily or consistent exposure. Damaged cells would develop differently and would represent such if compared in complete studies using EMF and chemical comparisons.**

Radiation Exposure: Studies of radiation exposure from patients receiving radiation therapy have an increased risk of developing NHL later in life. **This statement further implicates EMF radiation as more of a factor in developing NHL than otherwise accepted.**

Having a Weakened Immune System: People with weakened immune systems have an increased risk for NHL. **EMF intensity, depending on exposure, weakens immune systems constantly over time.**

Autoimmune Diseases: Diseases such as Rheumatoid arthritis have been linked to NHL. **EMF exposure would enhance the development of Rheumatoid Arthritis.**

CORRELATIONS BETWEEN RISK FACTORS FOR NHL AND FINDINGS IN THIS TEXT

Infections that directly transform lymphocytes: Some viruses can directly affect the DNA of lymphocytes, helping to transform them into cancer cells. **Long term high intensity EMF exposure has been shown to break down molecular structure and disrupt the structure of genes.**

Infections that cause chronic immune stimulation: Some long term infections may increase a person's risk of lymphoma by forcing their immune system to be constantly active. As more lymphocytes are made to fight infection, there is a greater chance for mutations in key genes to occur, which might eventually lead to lymphoma. Some of the lymphomas linked with these infections actually get better when the infection is treated. **Daily EMF exposure can cause more mutations and allow the cycle of a constant reoccurring infection to continue more aggressively.**

Final Findings and Comments

According to a report by two researchers from the University of Nebraska-Lincoln research testing found that the High Plains and California based Central Valley aquifer had concentration levels of uranium that tested 180 times the maximum recommended by the EPA. Nitrogen levels were tested to be 34 times higher than maximum concentration levels recommended. The two aquifers irrigated cropland that accounts for one-sixth of the annual revenue generated by US agriculture/ from the report. This report was just one of many examples found not only in these aquifers, but many other areas of the United States having radiation contaminated water. The areas reported involved an estimated 6 million people living in the contaminated areas tested. EMF/ Radiation contamination stretches from coast to coast and from state to state. Even in Hawaii, Alaska, and Canada it is of great concern. California seemed to have the greatest potential level of contamination per population numbers affected, that I found.

California has a strong history of radiation tainted water, through much of the state. Many Californians are being hit with a double radiation concern at the same time. An example of this would be radiation damage caused by EMF and EMF ionized water mixed with the concentrations of irradiated water from uranium contamination. A deadly cancer causing problem made more evident because there may be at least two different sources of radiation, one could be subjected to at the same time. This should be especially considered because of the amount of radium and radon contamination in the state making it more probable to take place.

I watched many videos and read about individuals that blamed roundup for their cancer and planned to sue Bayer. I also took random names of plaintiffs scheduled for future lawsuits against Bayer's Roundup. All of them had lived or were living in areas with reported high levels of some form of radiation contamination. I even was able to correlate a link to the first plaintiff I randomly found concerning Bayer's/Johnson and Johnson's baby powder. The person had won a multi million dollar judgment against Bayer. I then found areas where the person lived had magnetic field/ radiation contamination. When I continued to look, I found that in the same area the plaintiff had resided, traces of asbestos found in well water, was common. With the combination of the two components affecting the results being exposed in the environment, it should have been hard to prove Johnson and Johnson was at fault. EMF/ Radiation could have caused all of the types of the reaction, even the movement of the traces of Baby powder or any source of asbestos toward the ovaries. EMF/Radiation would give more credence to the reason cancer was gotten in the first place. If EMF/Radiation exposure was not considered Bayer, or any other entity, would always be found libel to some degree.

Some of the video I saw were of farms in Iowa believing

problems with their animals were caused by Roundup. They seemed to be very reputable farmers and professionals. When I researched I found several reports about contaminated wells throughout Iowa. One report said over 70% of private wells, in Iowa, were estimated to be contaminated. When farming years ago, there were times when my animals exhibited similar symptoms and examples, as stated in the video. The problem was not found to be any chemical, but toxins and poorly balanced formulations by Nutritionists' that were supposed to be experts in their field. Did they check their farms for "all" forms of contaminants.

In a case involving a Livermore California couple, the jury issued a 2 Billion dollar verdict against Bayer. It was stated that the couple sprayed roundup on three properties over many years in the Livermore area. When I looked back in history I found the Livermore area had numerous examples of radiation exposure. A nuclear toxic clean up site was noted because of nuclear open air dirty bomb testing in the1960's, just outside of Livermore. According to one report it affected the whole San Francisco area with radiation poisoning for many years. Clean up of the site was not done until the 1990's. Depleted uranium was found again in 2014 lying on the ground in many places at the site. What is not known is the effect of control burns at the site and close by. The smoke and ash could harm all that came in contact with it, years later. It was uranium 238. I also read reports of uranium contaminated groundwater in the Livermore area. Were these facts ever mentioned in the suit?

Reports were read and video was watched of individuals in other countries blaming Roundup for their problems. Every one of them had contamination factors that could easily justify their problems. None of those types of in field contamination were mentioned in the trial, or transcripts that I saw. The suits all directed to scientific "in lab findings", when basing causation and testing with no in- lab or out-lab allowance for EMF/Radiation exposure.

When viewing video taken of labs that do animal testing, many of the laboratories had lab animals being tested, stacked on top of each other, in housing containers. They looked very professional and clean. The first thing that came to my mind was how much they resembled my example of the aisles in stores. (*Image Index 9*) This time the scientist was walking down the center with the animals in the racks on the sides. Were the racks tested for positive ion exposure as I found to be true in stores? Could the animals and their water be bombarded daily by positive ions of a magnetic field during testing? Would the lab tech have minimal exposure, but not the animals? Testing would show whether that was true or not. EMF testing may be the only way to establish if Epigenetic and Genetic testing was accurate or not. EMF exposure would have to be compared and the accuracy of Epigenetic testing determined by measuring exposure levels. Without daily EMF ionization intensity testing, results should be considered incomplete. EMF positive ion intensity exposure could change just by housing unit location movements. How the housing containers are positioned, located, and time maximum intensity is absorbed could affect results in a positive or negative way determined by the lab researchers. Just simply moving the test animal a few inches from a magnetic field, each day, would make a difference. Individual mice/animals and groups of mice/animals being tested should have equal exposure to EMF ionization. It would be easy to skew results of a test, one way or the other, through changes in EMF positive ion exposure, especially if testing mice lasted a year or more. By increasing positive ion exposure, cells would be disrupted, and DNA structure could be harmed. Breakdown of any chemical in the study, without magnetic intensity monitoring, would not confirm if EMF was responsible. The chemical then would be the villain and the true villain, EMF exposure, would continue to create false results. EMF inconsistency during testing should be noted daily. EMF exposure genetic response may not show up for several generations of mice. The intensities of daily exposure would not be consistent so the effect would probably show inconsistent effects on the Genome of the mice. This inconsistent effect would be affected up or down by exposure intensity and the strength of the field. It would affect each mouse differently to the point that some mice living within inches of other mice would develop a different change to its genome than its neighbor. The test would seem to be inconsistent, but would only be mimicking the property's it was exposed to by the EMF. Even if a room could be made that magnetic intensity levels could be regulated through out testing, results would not be the same as in the field. It is like what was said earlier in the text, referring to a child losing his tooth, as well as, the carpenter's daily contact with various forms and intensities of radiation exposure.

What about the tubes of chemical, petri dishes, and contact/exposure mixes' being tested, over a period of time? Was their environment EMF/Radiation tested during the experiments? Many times cell tissue is kept in a refrigerator. Was the inside of the refrigerator tested for a magnetic intensity higher than the average of the earth, at that location? Some testing says Glyphosate reduces beneficial bacteria in the gut. Ionic exposure from EMF/Radiation can affect enzymes in bacteria and kill it. This is done by altering the hydrogen ions in the gut, making it more acidic. Has any documentation shown that EMF/Radiation testing was done while monitoring levels of gut bacteria, in lab animals that might alter genes? Ion radiation contamination in the gut could come from the type of water system, storage tanks, distribution lines, ground wells, and other sources already mentioned.

It has now been proven animals are showing us there is reason to avoid magnetic field ionization/radiation intensity.

Lower readings than we ever thought possible in the past, are affecting them.

So who can you believe?

There were a multitude of other claims against Bayer that could easily be linked to what was expressed in this text as the probable cause for the cancer diagnosis. If EMF and other radiation linked exposures are not considered, the testing should be invalid / not complete and lawsuits dropped. EMF/Radiation exposure at harmful levels in the United States is just too common. If I am correct on my findings, I can only imagine the harm that will come to our children and grandchildren through Epigenetic modifications, caused by types of radiation and corrections delayed by blaming a chemical needed for our assured food supply.

Convert your iPhone to a Magnetometer

Go to the app store on your iPhone. Search for Magnetometer and choose the app you prefer. I recommend Tesla-Bot. Click on x10 on right side of phone after app is applied. Turn up volume so you can follow. Your phone will sound like and act like a Geiger counter. The average magnetic induction of the earth here is about 46μT (micro tesla). The magnetic reading varies with temperature, sun, and moon phases as stated in text.

A Geiger counter or other instruments would need to be used for testing uranium, radium and radon levels.

Use your magnetometer to test the ability of trap/devices before you make a purchase. What you learn will allow you to make a better decision as to what limitations the device may have. That will give you more clarity. If the trap has a high internal variance when set, and even if other trappers say it is "the best trap available", you probably should keep on looking. You may want to test a larger size, different design, or another manufacturer. The magnetometer always told me accurately how some animals would respond as described in this text. It was the only way I found that every circumstance best addressed using a trap could be made with confidence. Remember, all traps tested caught animals. The devices that passed testing just caught many times more animals, faster, over the life of their use.

Image Index

Image 1

Image 2

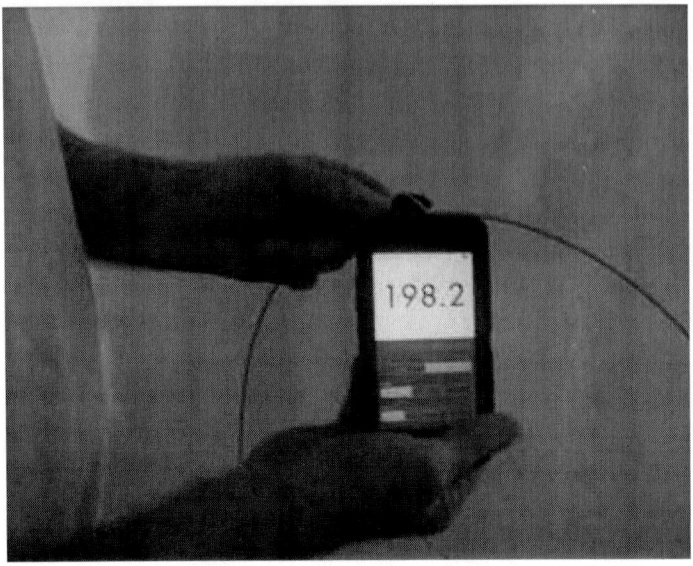

Image 3

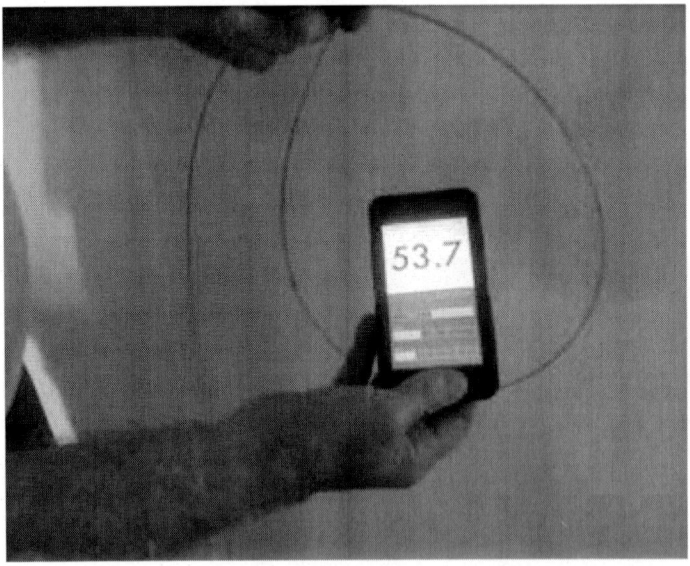

Image 4

Image 5

Image 6

Image 7

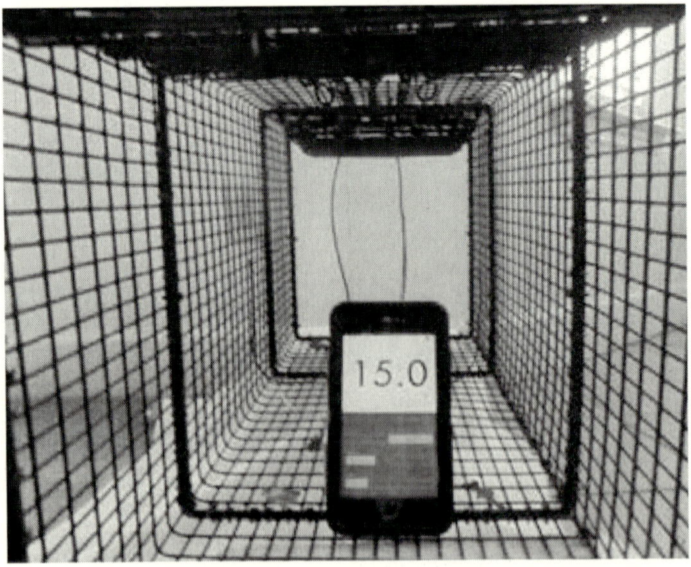

Image 8

Image 9

Made in United States
North Haven, CT
10 November 2026

22096551R00039